SURVEYIN
MA

An original Book by

Jim Crume P.L.S., M.S., CFedS

Co-Authors

Cindy Crume

Bridget Crume

Troy Ray R.L.S.

Mark Sandwick L.S.I.T.

PRINTED EDITION

*****

PUBLISHED BY:

Jim Crume P.L.S., M.S., CFedS

Parcel Boundary

Book 5 of this Math-Series

First publication: November, 2013

Printed by CreateSpace

Available on Kindle and other devices

TERMS AND CONDITIONS

Table of Contents

INTRODUCTION

Straight forward Step-by-Step instructions.

This book is just one part in a series of digital and printed editions on Surveying Mathematics Made Simple. The subject matter in this book will utilize the methods and formulas that are covered in the books that precede it. If you have not read the preceding books, you are encouraged to review a copy before proceeding forward with this book.

For a list of books in this series, please visit:

http://www.cc4w.net/ebooks.html

Prerequisites for this book:

A basic knowledge of geometry, algebra and trigonometry is required for the explanations shown in this book.

Book 1 - **Bearings and Azimuths** - How to add bearings and angles, subtract between bearings, convert from degrees-minutes-seconds to decimal degrees, convert from decimal degrees to degrees-minutes-seconds, convert from bearings to azimuths and convert from azimuths to bearings.

Book 2 - **Create Rectangular Coordinates** - How to calculate the northing and easting of an end point given the coordinates of the beginning point, bearing and distance of a line.

Book 3 - **Inverse Between Rectangular Coordinates** - How to determine the bearing and distance of a line given the coordinates for the beginning and ending point.

Book 4 - **Circular Curves** - How to calculate a circular curve, reverse curve, compound curve, Tangent In, Tangent Out and Local Tangent Bearing given only two parameters.

Definition: Parcel Boundary: A line of demarcation between adjoining parcels of land. A parcel boundary line can be described by a "metes and bounds" description, referral to a "lot and block" of a platted subdivision and defined by "aliquot parts" of the Public Land Survey System (PLSS).

PARCEL BOUNDARY

This book will build upon the lessons learned in the preceding books 1 through 4. It is important that you thoroughly understand the subject matter contained in the preceding books.

The processes that we will be going through will be to evaluate the legal description for a parcel of land, calculate coordinate pairs for each corner of the parcel, check the closure of the parcel and to calculate the area of the parcel.

LEGAL DESCRIPTION

Below is a legal description for a parcel of land that will be utilized for the lessons in this book.

Those portions of the Southwest quarter of Section 25, the Southeast quarter of Section 26 and the Northeast quarter of Section 35, Towhship 41 North, Range 30 East of the Gila and Salt River Meridian, Apache County, Arizona, described as follows:

Commencing at a found 3 1/2 inch Bureau of Land Management (BLM) Department of Interior 2006 brass cap marking the common corner for Sections 25, 26, 35 and 36 of said Township, from which the quarter corner for Sections 26 and 35 of said Township bears South 89°21'45" West, 2638.40 feet being marked with a 3 1/2 inch BLM 2006 brass cap;

Thence South 89°21'45" West, 305.47 feet along the south line of said Section 26 to a point on the northwesterly right of way line of route U.S. 160 (Tuba City - Four Corners) being the POINT OF BEGINNING;

Thence South 34°09'26" West, 47.03 feet along said northwesterly right of way line;

Thence North 55°52'03" West, 599.82 feet;

Thence North 34°09'22" East 600.24 feet;

Thence South 55°49'38" East 596.45 feet to a point that lies on said northwesterly right of way line;

Thence along said northwesterly right of way line, from a local tangent bearing of South 32°26'35" West, along a curve to the right, having a radius of 7539.44 feet, a curve length of 225.59 feet, through a central angle of 1°42'52" to the point of tangency;

Thence South 34°09'26" West, 327.23 feet along said northwesterly right of way line to the POINT OF BEGINNING.

Said parcel of land containing 359,655 square feet, more or less.

The legal description will need to be deciphered by extracting from the course for each boundary line for it's bearing and distance. These bearings and distances will be used to calculate rectangular coordinates.

The legal description contains qualifiers that would be taken into consideration during an actual boundary line retracement. This book is only going to concentrate on the mathematical portion of the legal description.

We will start at the commencing point.

Commencing at a found 3 1/2 inch Bureau of Land Management (BLM) Department of Interior 2006 brass cap marking the common corner for Sections 25, 26, 35 and 36 of said Township, from which the quarter corner for Sections 26 and 35 of said Township bears South 89°21'45" West, 2638.40 feet being marked with a 3 1/2 inch BLM 2006 brass cap;

Thence ***South 89°21'45" West, 305.47*** *feet along the south line of said Section 26 to a point on the northwesterly right of way line of route U.S. 160 (Tuba City - Four Corners) being the POINT OF BEGINNING;*

The first course is: S89°21'45"W 305.47' which will take us to the Point of Beginning (POB).

Thence ***South 34°09'26" West, 47.03*** *feet along said northwesterly right of way line;*

The second course is: S34°09'26"W 47.03'.

Thence ***North 55°52'03" West, 599.82*** *feet;*

The third course is: N55°52'03"W 599.82'.

> *Thence* ***North 34°09'22" East 600.24*** *feet;*

The fourth course is: N34°09'22"E 600.24'.

> *Thence* ***South 55°49'38" East 596.45*** *feet to a point that lies on said northwesterly right of way line;*

The fifth course is: S55°49'38"E 596.45'.

> *Thence along said northwesterly right of way line, from a local tangent bearing of* ***South 32°26'35" West****, along a curve to the right, having a radius of* ***7539.44*** *feet, a curve length of 225.59 feet, through a central angle of* ***1°42'52"*** *to the point of tangency;*

The sixth course takes us through a curved segment starting with a Local Tangent Bearing. We need to convert the curve information into the Chord Bearing and Distance that we can then use to calculate the rectangular coordinates.

From the information shown in the legal description, we are given the following:

Δ = 1°42'52"

R = 7539.44

L.T.B. In = S32°26'35"W

From this information we can calculate the Chord Bearing and Distance (C).

Chord Bearing = L.T.B. In (+/-) (Δ / 2)

Chord Bearing = S32°26'35"W + (1°42'52" / 2)

Chord Bearing = **S33°18'01"W**

C = 2 * R * Sin(Δ / 2)

C = 2 * 7539.44 * Sin(1°42'52" / 2)

C = **225.59199**

Note: The formulas shown above were taken from Book 4 "Circular Curves".

The sixth course is: S33°18'01"W 225.59 (Along the chord).

> *Thence* ***South 34°09'26" West, 327.23*** *feet along said northwesterly right of way line to the POINT OF BEGINNING.*

The final course is S34°09'26"W 327.23' back to the Point of Beginning.

Figure 1 shows the courses for this parcel.

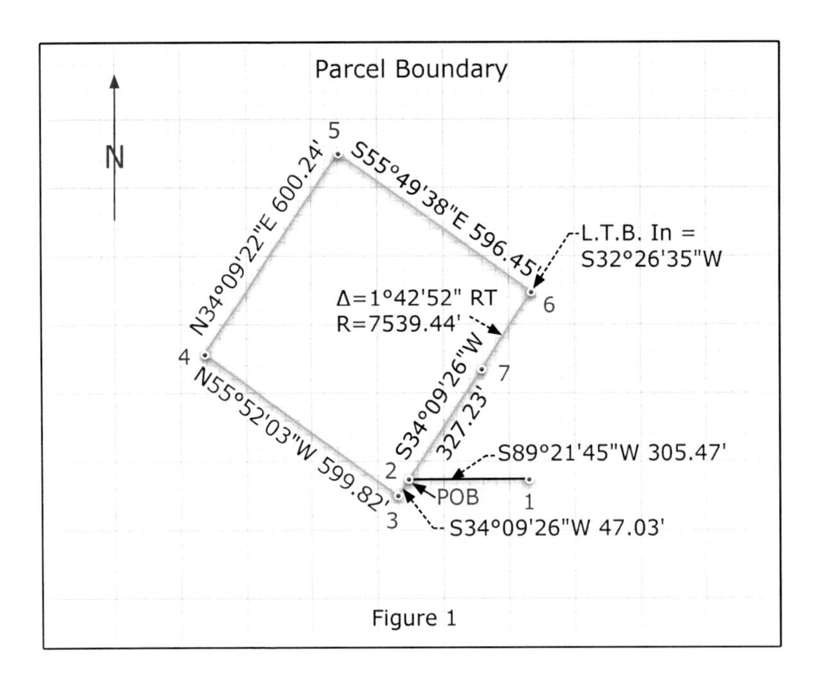

Figure 1

Place the course Bearings and Distances in a table or spreadsheet. Convert the Bearings to Azimuths in decimal degrees format. See the formula in Book 1 "Bearings and Azimuths" on how to convert to Azimuths and Decimal Degrees.

Note: The following charts were generated by a spreadsheet program. The calculated numbers that are shown in these charts are double precision floating point values. Values generated by a hand held calculator may be slightly different than though's shown below and should be insignificant.

Pt. No.	Bearing - DMS	Azimuth - D.ddd	Distance	Desc.
1				Sec Cor
	S89°21'45"W	269.36250000	305.47	Lead in
2				POB
	S34°09'26"W	214.15722222	47.03	
3				
	N55°52'03"W	304.13250000	599.82	
4				
	N34°09'22"E	34.15611111	600.24	
5				
	S55°49'38"E	124.17277778	596.45	
6				
	S33°18'01"W	213.30027778	225.59	Chord
7				
	S34°09'26"W	214.15722222	327.23	
2a				POB

The next step is to calculate the Latitude (Lat) and Departure (Dep) for each course.

Use the following formulas to calculate the Lat and Dep for each course:

Lat = Cos(Azimuth) * distance

Dep = Sin(Azimuth) * distance

Note: The formulas shown above were taken from Book 2 "Create Rectangular Coordinates".

Solution for course between Points 1 and 2:

Lat = Cos(269.36250000) * 305.47

Lat = **-3.39873**

Dep = Sin(269.36250000) * 305.47

Dep = **-305.45109**

Repeat this process for the remaining courses. The following table shows the solutions for the Lat and Dep for all courses. It is important to include the algebraic sign.

Pt. No.	Latitude	Departure
1	Cos(Az)*D	Sin(Az)*D
	-3.39873	-305.45109
2		
	-38.91733	-26.40573
3		
	336.56416	-496.49632
4		
	496.70514	337.00455
5		
	-335.02021	493.47144
6		
	-188.54918	-123.85497
7		
	-270.78283	-183.72842
2a		

The next step is to check the Error of Closure for the legal description. Add the Lat column and the Dep column starting from the Point of Beginning then continuing with each course. The summation of the Lat column and the Dep column should equal zero.

For this example, add the Lat and Dep values from Points 2 to 2a as highlighted in green.

Note: Only add the Lat and Dep values from the Point of Beginning then around the parcel and back to the Point of Beginning. DO NOT include any lead in courses.

$Lat_{(ERROR)} = \sum(Lat_N + Lat_{N+1...})$

$Lat_{(ERROR)} = \sum(-38.91733 + 336.56416 + ...)$

$Lat_{(ERROR)} =$ **-0.00025**

$Dep_{(ERROR)} = \sum(Dep_N + Dep_{N+1...})$

$Dep_{(ERROR)} = \sum(-26.40573 + -496.49632 + ...)$

$Dep_{(ERROR)} =$ **-0.00945**

Note: It is common for there to be small error of closures for closed parcels due to the rounding of bearings to the nearest second and distances to two decimal places. Larger error of closures indicate that there is a discrepancy in one or more of the courses of the legal description. It is beyond the scope of this book to define the processes of locating where the discrepancy is and the methods for correcting the discrepancy.

The next step is to create Rectangular Coordinates for each point.

Assign the following ground coordinates for Point 1:

Point 1

Northing = 2157691.70913

Easting = 1016642.48026

Note: The starting point can be assumed coordinates or pre-defined coordinates from a prior survey or State Plane

Coordinates that have been modified to ground coordinates utilizing a Grid Adjustment Factor.

Point 2 (POB)

Northing = Northing Point 1 + Lat (between Points 1 & 2)

Northing = 2157691.70913 + (-3.39873)

Northing = **2157688.31040**

Easting = Easting Point 1 + Dep (between Points 1 & 2)

Easting = 1016642.48026 + (-305.45109)

Easting = **1016337.02917**

Repeat this process for the remaining courses. The following table shows the solutions for the Northing and Easting for all courses. It is important to include the algebraic sign when adding the Lat and Dep columns.

Pt. No.	Northing	Easting
1	2157691.70913	1016642.48026
2	2157688.31040	1016337.02917
3	2157649.39307	1016310.62344
4	2157985.95724	1015814.12712
5	2158482.66237	1016151.13166
6	2158147.64216	1016644.60310
7	2157959.09298	1016520.74813
2a	2157688.31015	1016337.01970

Note: The difference between Points 2a and 2 for the Northing and Easting should be the same as the Error of Closure for the Lat and Dep. If they're not, then recheck your math.

Northing difference = Northing Point 2a - Northing Point 2

Northing difference = 2157688.31015 - 2157688.31040

Northing difference = **-0.00025** ($Lat_{(ERROR)}$ - 0.00025)

Easting difference = Easting Point 2a - Easting Point 2

Easting difference = 1016337.01973 - 1016337.02917

Easting difference = **-0.00944** ($Dep_{(ERROR)}$ - 0.00945)

Note: As you can see from the check values above that even at 5 decimal places rounding error will occur. This is normal and is negligible for survey work.

NOTES

PARCEL AREA

There are basically two methods of calculating area's of a parcel. One is the "Double Meridian Distance" and the other is "Cross Coordinate" method.

The easiest of the two methods is the "Cross Coordinate" method which we will utilize for the preceding parcel.

The first thing we want to do is modify the coordinates that were previously calculated in order to reduce the number of places to make it easier to perform the multiplication with a calculator.

Modify the North coordinate by subtracting 2150000 from each coordinate and modify the East coordinate by subtracting 1010000 from each coordinate.

Below are the modified values:

Pt. No.	Northing	Easting
1	----	----
2	7688.31040	6337.02917
3	7649.39307	6310.62344
4	7985.95724	5814.12712
5	8482.66237	6151.13166
6	8147.64216	6644.60310
7	7959.09298	6520.74813
2a	7688.31040	6337.02917

Points number 2 through 2a are the coordinate values for the closed parcel. Point number 1 will not be used. Since we need a perfectly closed parcel before calculating the area, Point 2a needs to equal Point 2 thus reducing the coordinate rounding error.

NOTES

CROSS COORDINATE METHOD

Start by multiplying the north coordinate of Pt 2 by the east coordinate of Pt 3 and place this value in the Product 1 column.

7688.31040 * 6310.62344 = **48518031.824**

Next multiply the north coordinare of Pt 3 by the east coordinate of Pt 4 and place this value in the Product 1 column.

7649.39307 * 5814.12712 = **44474543.700**

Repeat this process for the remaining points in the list. See the black arrows in the table below.

Next multiply the north coordinate of Pt 3 by the east coordinate of Pt. 2 and place this value in the Product 2 column.

7649.39307 * 6337.02917 = **48474427.070**

Next multiply the north coordinate of Pt 4 by the east coordinate of Pt 3 and place this value in the Product 2 column.

7985.95724 * 6310.62344 = **50396368.950**

Repeat this process for the remaining points in the list. See the brown arrows in the table below.

Pt. No.	Northing	Easting	Product 1	Product 2
1	----	----		
2	7688.31040	6337.02917		
3	7649.39307	6310.62344	48518031.767	48474427.010
4	7985.95724	5814.12712	44474543.674	50396368.887
5	8482.66237	6151.13166	49122674.407	49319277.314
6	8147.64216	6644.60310	56363924.702	50117219.671
7	7959.09298	6520.74813	53128722.377	52885013.885
2a	7688.31040	6337.02917	50437004.357	50133535.632
		Total =	302044901.284	301325842.397
		Product 1 - Product 2 =	719058.887	
		divided by 2 =	359529.4435	Square Feet

Now total the columns for Product 1 & 2. Subtract these two values and divide by 2. This is the area of the parcel in square feet excluding the curved segment.

Almost there. This parcel contains a curved segment. To complete the area for this parcel we need to compute the area of the curved segment and add it to the above area.

Note: The curved segment area for a parcel will either be added or subtracted from the area by cross coordinate method area depending upon the direction of the curve.

NOTES

AREA OF CURVED SEGMENT

The curved segment lies between Pt 6 and 7. The curve information given in the legal description is as follows:

$\Delta = 1^{o}42'52''$

$R = 7539.44$

There is a two stage process to calculate the area of the curved segment. First we need to find the area to the arc and next the area to the chord. The difference between these two areas is the curved segment area.

Area to the Arc:

Area to Arc = $(\Delta * \pi * R^2) / 360$

Area to Arc = $(1^{o}42'52'' * \pi * 7539.44^2) / 360$

Area to Arc = **850450.356** Sq. Ft.

Area to the Chord:

Area to Chord = $(R^2 * \mathrm{Sin}(\Delta)) / 2$

Area to Chord = $(7539.44^2 * \mathrm{Sin}(1^{o}42'52'')) / 2$

Area to Chord = **850323.451** Sq. Ft.

Segment Area:

Segment Area = Area to Arc - Area to Chord

Segment Area = 850450.356 - 850323.451

Segment Area = **126.905** Sq. Ft.

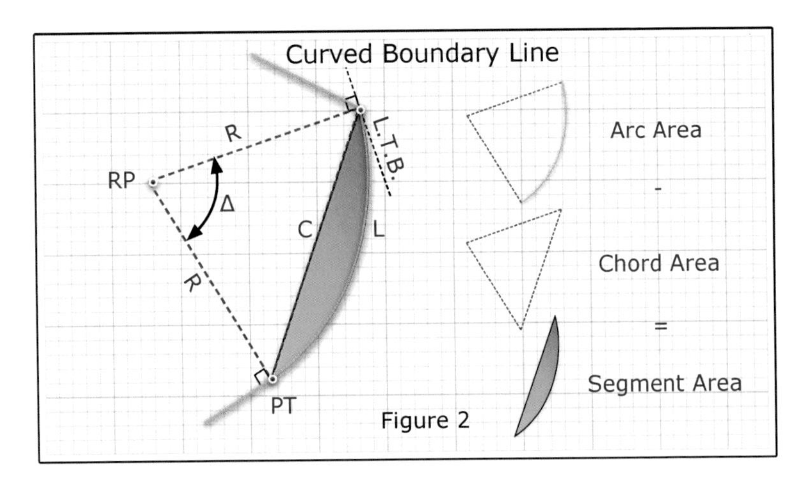

Figure 2

Final parcel area = Cross Coordinate Area + Segment Area

Final Parcel Area = 359529.444 + 126.905

Final Parcel Area = **359656.349** Sq. Ft. (Round to the nearest square foot: 359656 Sq. Ft.)

Final Parcel Area in Acres = 359656.349 / 43560

Final Parcel Area in Acres = **8.257**

The area shown in the Legal Description is 359,655 Sq. Ft. The final calculated area is within 1 Sq. Ft. of the legal description which indicates that the computer generated area versus the manually calculated area are utilizing different decimal precision. This rounding error is unavoidable and is negligible.

Computer generated areas use double precision floating point through out all of the calculations. You are limited to the number of decimal places on a calculator therefore using a calculator is less accurate. Where you decide to round numbers throughout the calculation process will also have an affect on the final number.

There are different schools of thought among survey professionals when it comes to rounding the area in square feet to a whole number. I personally round all parcel areas to the nearest square foot due to the rounding errors that do occur depending on the method that is used to calculate the area and I feel that the preceding mathematical analysis proves that hypotheses.

The bottom line is that it is impossible to calculate the area of any parcel to an exact square footage or to duplicate the method that someone else has used to calculate an area. To suggest that it can be calculated to the nearest tenth, hundredth or thousandth of a square foot is highly arguable and proven heretofore. When it comes to analyzing a legal description, the area is on the bottom of the list for the “Order of precedence”.

NOTES

ABOUT THE AUTHOR

Jim Crume P.L.S., M.S., CFedS

My land surveying career began several decades ago while attending Albuquerque Technical Vocational Institute in New Mexico and has traversed many states such as Alaska, Arizona, Utah and Wyoming. I am a Professional Land Surveyor in Arizona, Utah and Wyoming. I am an appointed United States Mineral Surveyor and a Bureau of Land Management (BLM) Certified Federal Surveyor. I have many years of computer programming experience related to surveying.

This book is dedicated to the many individuals that have helped shape my career. Especially my wife Cindy. She has been my biggest supporter. She has been my instrument person, accountant, advisor and my best friend. Without her, I would not be the professional I am today. Cindy, thank you very much.

Other titles by this author:

http://www.cc4w.net/ebooks.html

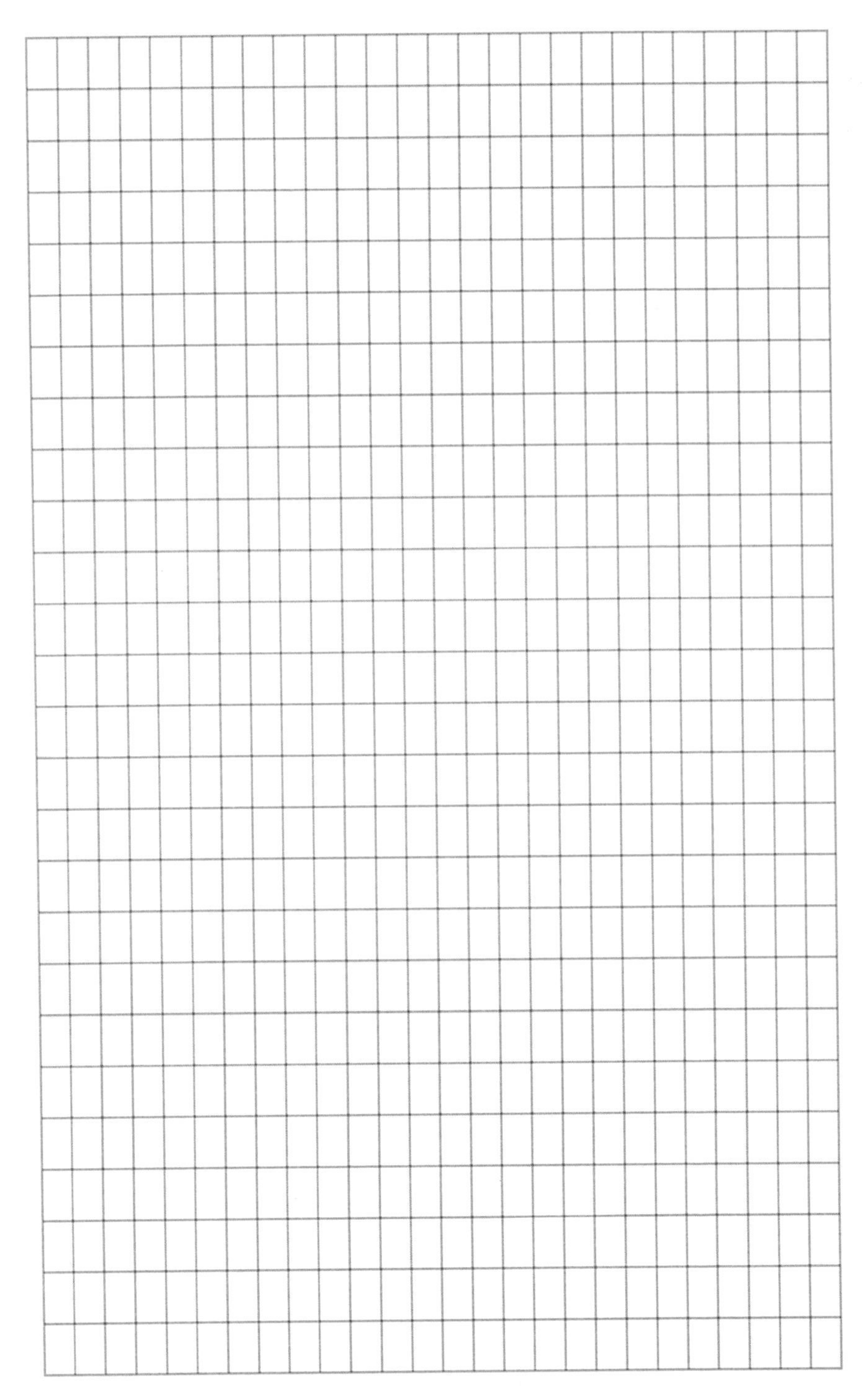

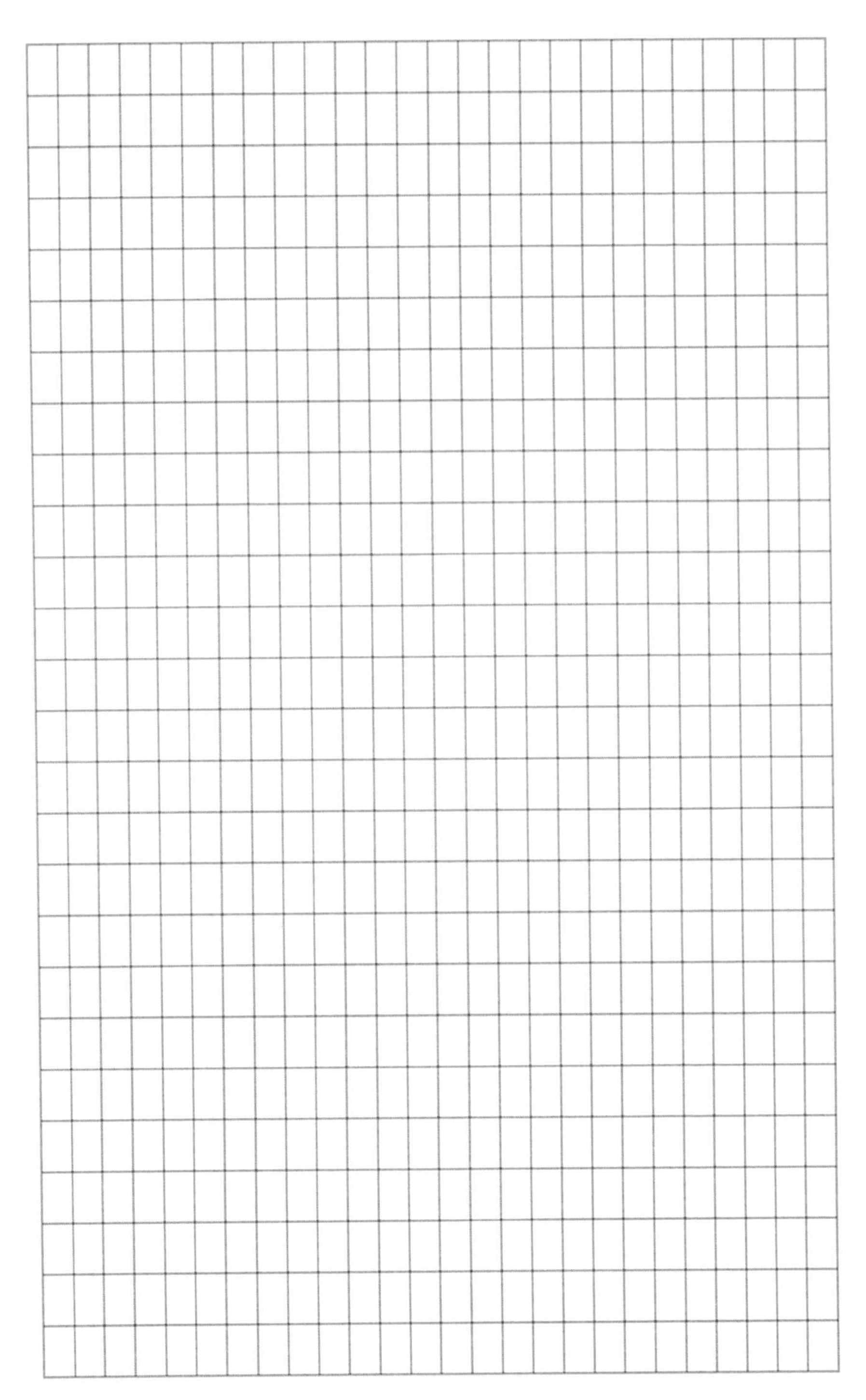

Made in the USA
San Bernardino, CA
13 February 2019